PB98-917007
NTSB/HZM-98/01/SUM

# NATIONAL TRANSPORTATION SAFETY BOARD

WASHINGTON, D.C. 20594

## HAZARDOUS MATERIALS ACCIDENT SUMMARY REPORT

FAILURE OF TANK CAR TEAX 3417 AND SUBSEQUENT
RELEASE OF LIQUEFIED PETROLEUM GAS
PASADENA, TEXAS, NOVEMBER 22, 1997

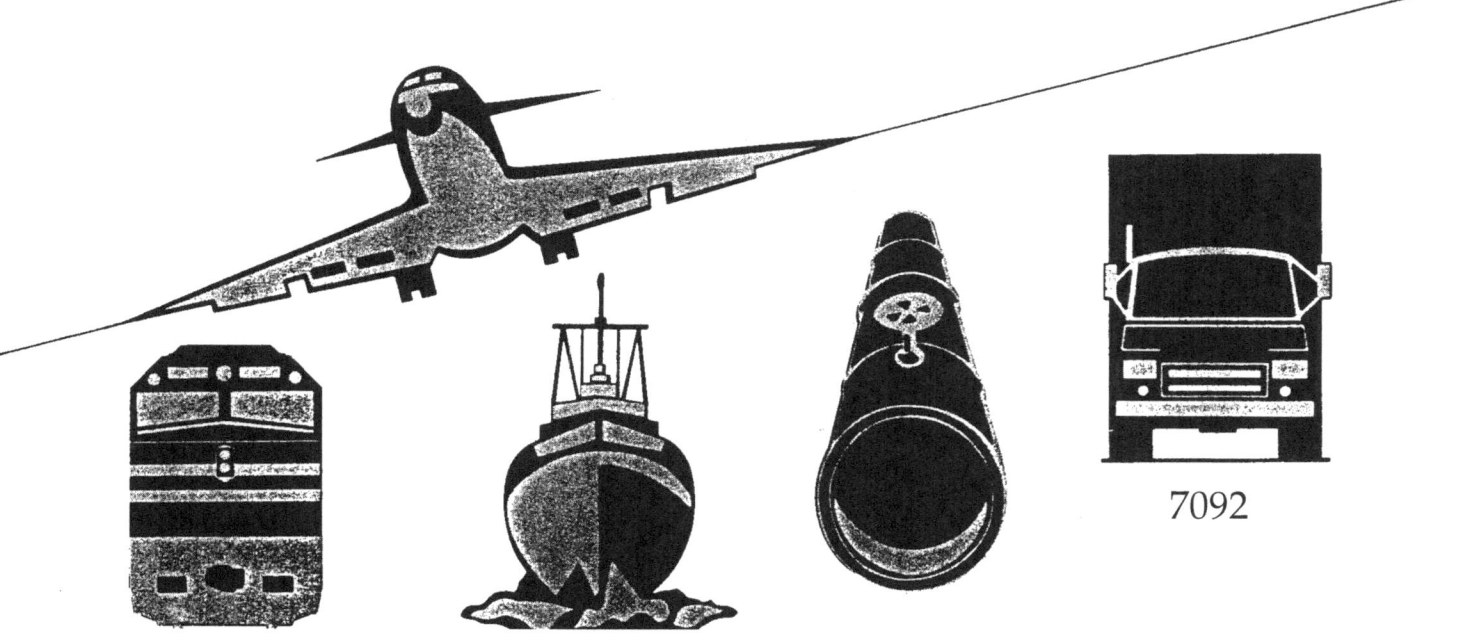

7092

**Abstract:** On November 22, 1997, a frost ring that signified product leakage was discovered on the bottom center of a tank car that was being unloaded at the Georgia Gulf Corporation chemical plant in Pasadena, Texas. The tank car contained 29,054 gallons of a propylene/propane mixture, a liquefied flammable gas. The tank car had been purged with cryogenic nitrogen on October 17, about a month before the accident. No injuries or fatalities were reported as a result of the failure of the tank car. Georgia Gulf estimated that approximately 52 gallons of the cargo were released. Total damage, including the cost of the clean up, loss of product, and repair of the tank car, was estimated to be slightly less than $9,300.

The safety issues discussed in this report are the need to safeguard tank cars adequately when they are being purged with nitrogen and the use of engineering analyses of the properties of tank car steels in the development of industry-recommended procedures for the purging of tank cars with nitrogen.

As a result of its investigation, the National Transportation Safety Board issued recommendations to the Compressed Gas Association, Inc., the Federal Railroad Administration, and the Association of American Railroads.

The National Transportation Safety Board is an independent Federal agency dedicated to promoting aviation, railroad, highway, marine, pipeline, and hazardous materials safety. Established in 1967, the agency is mandated by Congress through the Independent Safety Board Act of 1974 to investigate transportation accidents, determine the probable causes of the accidents, issue safety recommendations, study transportation safety issues, and evaluate the safety effectiveness of government agencies involved in transportation. The Safety Board makes public its actions and decisions through accident reports, safety studies, special investigation reports, safety recommendations, and statistical reviews.

Recent publications are available in their entirety on the Web at http://www.ntsb.gov/. Other information about available publications may be obtained from the Web site or by contacting:

**National Transportation Safety Board**
**Public Inquiries Section, RE-51**
**490 L'Enfant Plaza, S.W.**
**Washington, D.C. 20594**
**(202) 314-6551**
**(800) 877-6799**

Safety Board publications may be purchased, by individual copy or by subscription, from:

**National Technical Information Service**
**5285 Port Royal Road**
**Springfield, Virginia 22161**
**(703) 605-6000**

# Hazardous Materials Accident Summary Report

Failure of Tank Car TEAX 3417 and Subsequent Release of Liquefied Petroleum Gas, Pasadena, Texas, November 22, 1997

NTSB/HZM-98/01/SUM
PB98-917007

Notation 7092
Adopted: December 1, 1998

National Transportation Safety Board
490 L'Enfant Plaza East, S.W.
Washington, D.C. 20594

this page intentionally left blank

this page intentionally left blank

# Introduction

About 12:00 p.m., central standard time, on November 22, 1997, a frost ring that signified product leakage was discovered on the bottom center of tank car TEAX 3417 at the Georgia Gulf Corporation chemical plant in Pasadena, Texas. (See figure 1.) When the plant's employees discovered the leak, they were weighing the tank car, which contained 140,377 pounds (29,054 gallons) of a propylene/propane mixture, a liquefied flammable gas. The employees successfully offloaded most of the mixture from the tank car to fixed storage tanks. To transfer the residual cargo, they isolated the tank car. By 7:00 a.m. the next morning, November 23, they had transferred the residual to a cargo tank truck.

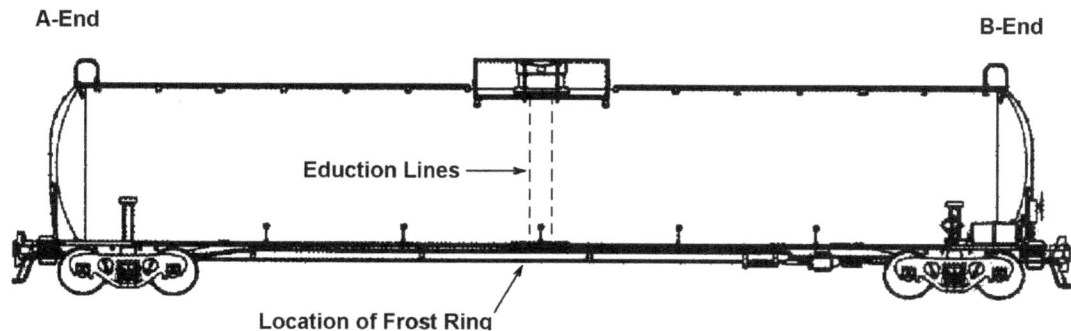

**Figure 1. Schematic view of typical tank car used for transportation of liquefied petroleum gas and anhydrous ammonia**

No injuries or fatalities were reported as a result of the failure of the tank car. Georgia Gulf estimated that approximately 250 pounds (52 gallons) of the cargo were released. Total damage, including the cost of the clean up, loss of product, and repair of the tank car, was estimated to be slightly less than $9,300.

A postaccident examination revealed that the product had been released through a circumferential crack in the bottom center of the tank. A metallurgical examination of the crack surfaces revealed indications of brittle (cleavage) fracture from a single event that had cooled the tank car to less than -50 °F. (The tank car had been purged with cryogenic nitrogen on October 17, about a month before the accident.)

Although the accident involved no injuries and only relatively minor damage, the shipping and transporting of liquefied flammable gases and nonflammable gases (such as anhydrous ammonia and chlorine) in pressure tank cars that may have such cracks in their tanks as tank car TEAX 3417 did significantly increase the likelihood of a more catastrophic failure and pose a threat to the public safety. Of the nearly 230,000 tank cars in the North American fleet, about 53,000 are used for the transportation of liquefied

flammable and nonflammable gases. Further, most of the 110 tank car manufacturing and repair facilities in the United States, Canada, and Mexico that are certified by the Association of American Railroads (AAR) have the capability to perform nitrogen purging of tank cars. Although the number of tank cars purged per year varies widely by tank car facility, nitrogen purging of tank cars is a commonly performed procedure that, without adequate safeguards, can potentially result in thermal shock failure of the tank.

# Preaccident Events

## Tank Car Information

Tank car TEAX 3417 (originally NATX 34899) was 1 of 100 tank cars (NATX 34839 through 34938) built in 1970 as U.S. Department of Transportation (DOT)[1] specification 112A340W tank cars. Trinity Industries, Inc., (Trinity) built the tank, and North American Car Corporation[2] constructed and assembled the tank car. The tank cars had a 33,900-gallon nominal capacity and were designed to carry liquefied petroleum gas (LPG) or anhydrous ammonia. The tank shell and heads were 0.603 inch thick and constructed of AAR specification TC-128 grade B steel.

In 1978, to comply with new amendments to the DOT *Hazardous Materials Regulations*[3] pertaining to tank cars transporting flammable gases, tank car NATX 34899 was converted from a DOT class 112A tank car to a DOT class 112J tank car. The conversion included installing thermal protection (1-inch mineral wool blankets), top and bottom shelf couplers (type "F"), and 0.5-inch steel head shields.[4]

## Inspection and Tests

Tank car TEAX 3417 arrived at the Bayou Railcar Services, Inc., (Bayou) repair shop in Holden, Louisiana, on October 1, 1997, for routine maintenance and testing. According to a test certificate, Bayou conducted a hydrostatic pressure test of the tank on

---

[1] See last page of this document for a list of all acronyms and abbreviations used in this report.

[2] North American Car Corporation is no longer in business. General Electric Railcar Services, Inc., now owns or operates much of the North American-built tank car fleet.

[3] Title 49 *Code of Federal Regulations* (CFR) 171-180.

[4] Safety recommendations issued by the National Transportation Safety Board from the 1970s through the mid-1990s provided the impetus to amend the regulations to require these safety features, not only on tank cars transporting liquefied flammable gases, but also on tank cars transporting a broader variety of hazardous materials, including nonflammable compressed gases, materials that are poisonous by inhalation, and specific environmentally harmful materials.

October 11 and a test of the safety relief valve on October 14.[5] The test certificate did not indicate any exceptions or abnormalities for either test. At the request of the owner of the tank car, Transportation Equipment, Inc., (TEI) of Houston, Texas, Bayou purged the tank car on October 17 with cryogenic nitrogen gas because of a change in the product to be carried. (The purging process involves injecting nitrogen gas through a tank car's liquid or vapor lines to displace residual vapors from previous cargoes.) Late that day, Bayou released the tank car, and it was shipped empty to Clark Refining and Marketing (Clark), the new lessee, located in Hartford, Illinois, a suburb of St. Louis, Missouri.

## Movements

The tank car arrived at Clark on October 28 and was loaded on November 6 with 140,377 pounds (29,054 gallons at a loading temperature of 41 °F) of a propylene/propane mixture. Although Clark employees were unable to recall loading tank car TEAX 3417 specifically, records show that the tank car was loaded from another tank car that was utilized for storage. Clark released tank car TEAX 3417 on November 7, and it was delivered to Georgia Gulf on the evening of November 21.

Investigators from the National Transportation Safety Board documented the movement of the tank car from the time that Bayou released it to the time that it arrived at Georgia Gulf and also obtained data about the temperatures of the ambient air to which the tank car had been exposed. On the trip from Bayou to the St. Louis area (October 17 to October 21), the tank car was not exposed to temperatures below 24 °F. Between October 21 and November 13, when the tank car was shipped to Pasadena, the lowest temperature that it was exposed to was 30 °F on November 12. During the movement of the tank car from St. Louis to Pasadena from November 13 to November 17, it was not exposed to temperatures below 22 °F. From November 17 to the accident date of November 22, local temperatures in the Houston area (including Pasadena) remained above 32 °F.

# The Accident

According to Georgia Gulf, on the morning of November 22, a plant employee discovered that the tank car had been delivered without placards and that the LPG stencil on the tank car had been obscured by paper and tape.[6] Consequently, sampling was necessary

---

[5]For the hydrostatic pressure test of the tank, a pressure of 340 pounds per square inch, gauge, (psig) was held for 20 minutes. The safety relief valve had a start-to-discharge pressure of 282 psig. Both tests complied with and satisfied DOT testing requirements.

[6]Under the DOT *Hazardous Materials Regulations* (49 CFR 171-180), each end and side of a tank car must have placards for a flammable compressed gas, and each side of the tank car must be stenciled with *Liquefied Petroleum Gas*. Clark maintained that when the tank car was shipped, it was properly placarded

to confirm the identity of the cargo. Approximately 12:00 p.m., as the tank car was weighed and the cargo sampled, plant employees observed a frost ring on the bottom center of the tank jacket. They offloaded most of the cargo following standard transfer procedures (injecting propylene/propane vapors into a tank car to increase its internal pressure so that the cargo flows to a lower pressure storage tank). However, increasing the internal pressure of tank car TEAX 3417 also increased the leakage rate. Consequently, at 4:10 p.m. Georgia Gulf stopped using standard transfer procedures and decided to transfer the remaining cargo to a cargo tank truck. A suitable cargo tank was located at approximately 3:00 a.m. November 23, and transfer of the residual cargo was completed by 7:00 a.m. that same day.

While plant personnel were attempting to offload the tank car, the plant implemented its emergency response plan. An incident command post was established, and by about 3 p.m. on November 22, a designated incident commander had assumed control of the response activities. Plant personnel serving as emergency responders utilized a fire monitoring system to apply a water fog to the tank car to keep the escaping vapors from reaching an explosive limit. Throughout the incident, vapor monitoring equipment was used to measure the concentration of propylene/propane vapors in the air. "Hot" and "safe" zones were established based on the percentage of the lower explosive limit[7] measured. At no point during the incident did the monitoring equipment detect a reading at or above the lower explosive limit. However, areas adjacent to the leaking tank car were evacuated as a precautionary measure.

On December 3, Georgia Gulf sought authorization from the Federal Railroad Administration (FRA) to move the damaged tank back to Bayou for repairs. The FRA granted approval on December 4, and the tank car arrived at Bayou on December 19.

# Postaccident Investigation

## Visual Examination

On January 7, 1998, Safety Board investigators and representatives of Bayou, the FRA, the AAR, Trinity, and the TEI, convened at Bayou to inspect the tank car visually. Before the inspection, Bayou had removed a portion of the tank jacket to provide an unobstructed external view of the bottom center of the tank. The exposed tank section contained a circumferentially oriented crack that was approximately 31 inches long. From the tank interior, the crack extended under one corner of the guide bracket for the A-end eduction pipe into the weld joining the bracket to the tank and through the tank wall

---

and stenciled.

[7]The lower explosive limit is the lowest concentration of the material in air that forms a flammable mixture.

directly below the eduction pipe. (See figure 2.) A smaller crack branched off from this crack at a location between the legs of the pipe guide bracket. From the exterior of the tank car wall, the two cracks appeared to be a single Y-shaped crack. (See figure 3.)

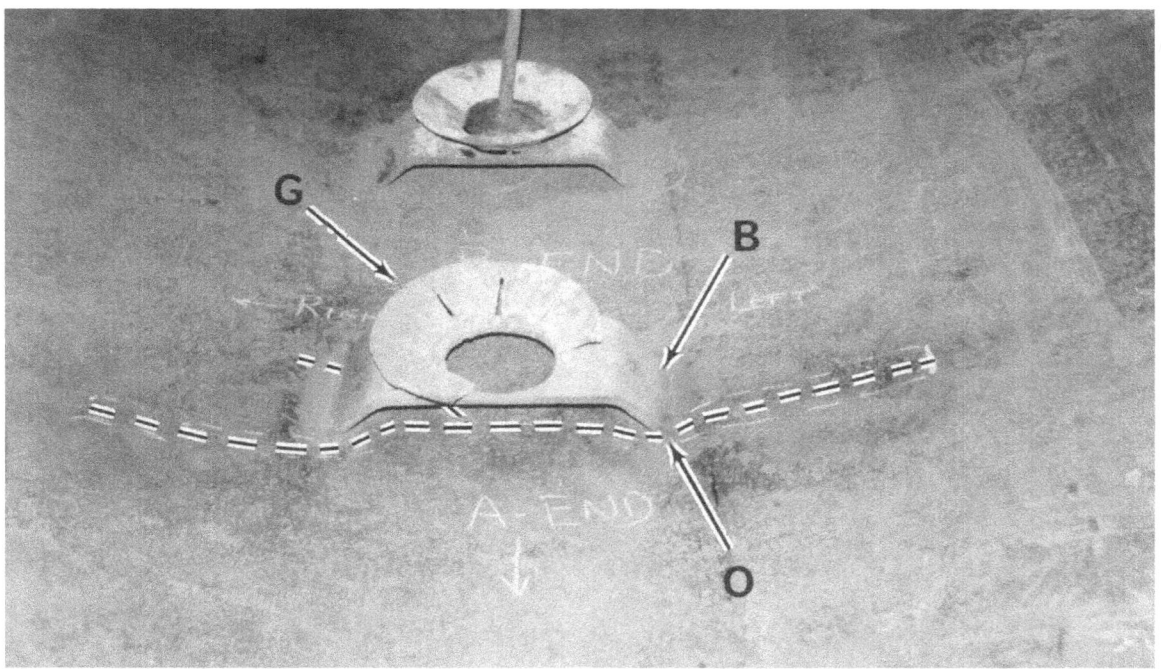

**Figure 2. Interior of tank car TEAX 3417**
(The A-end eduction pipe bracket *B* and guide *G* are in the foreground. The circumferential crack is indicated by the dotted line. The location of the crack origin is marked by the arrow *O*.)

Both the A-end and the B-end pipe guide brackets, which secure the bottom ends of the eduction pipes, were welded directly to the tank, a permissible practice at the time the tank was constructed. The A-end pipe guide, which was welded to the top of the pipe guide bracket and served to position the eduction pipe, did not match its B-end counterpart. The A-end guide contained a series of notches, whereas the B-end guide appeared as a solid section of metal. Additionally, about a fourth of the A-end guide was missing. Trinity later confirmed that the A-end guide had been modified after the original construction, whereas the B-end guide appeared to be original. An examination of the A-end eduction pipe, which had been removed from the tank before the January meeting, revealed a worn and eroded section at the bottom end whose orientation matched that of the broken section of the pipe guide. A walk-around inspection of the tank car revealed no visible damage to either the A- or the B-end striker plates, draw bars, or couplers.

The bottom section of the tank with the crack and the A-end pipe guide and bracket was cut out and sent to the Safety Board's materials laboratory for more detailed examination.

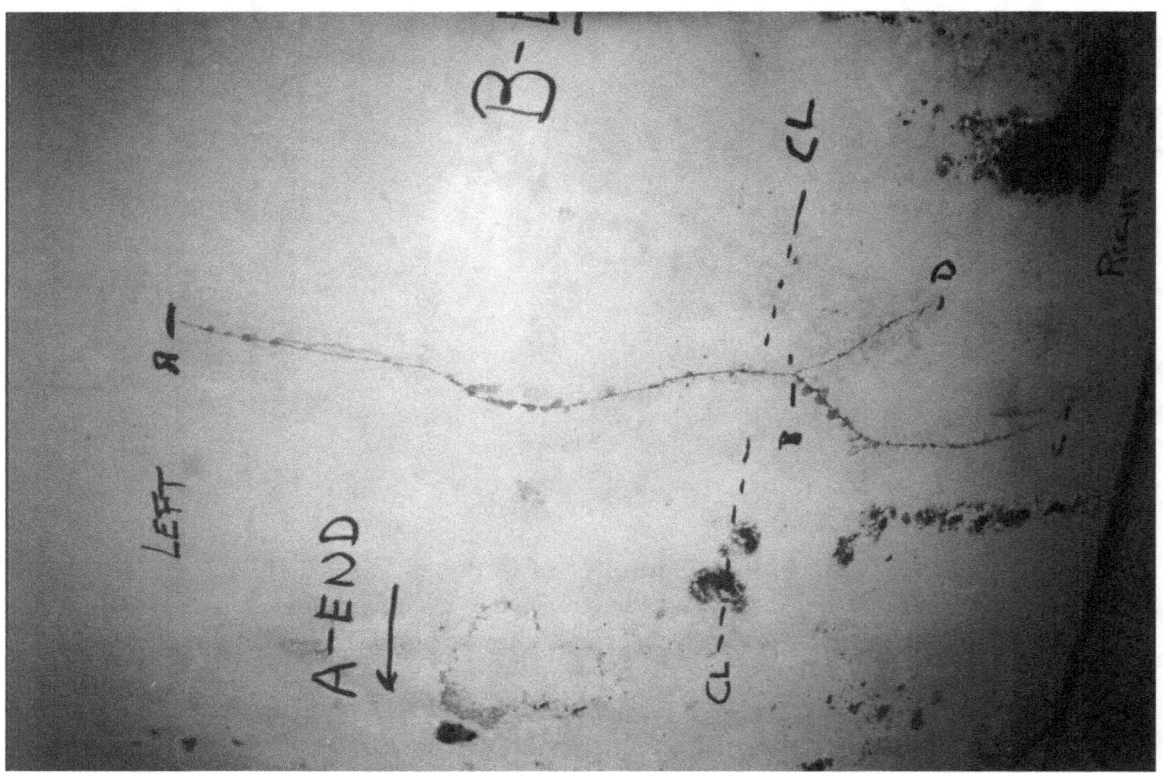

**Figure 3. Exterior view of tank car TEAX 3417 with view of circumferential crack at bottom center of tank car**

## Metallurgical Examination

At the Safety Board's materials laboratory, saw cuts were made on the section of the tank car wall to expose the 31-inch circumferential crack. A visual examination of the excised fracture revealed a river pattern[8] that originated from a void in the weld at one

_____

[8]An overstress fracture that contains features similar to nested letters *V*, where the points of the chevrons are traced back to the fracture origin.

corner of a bracket leg. The void was at the center of the weld joint between the bracket leg and the inside surface of the tank wall. The void was about 0.2 inch square and filled with black-gray porous material typical of welding slag. Based on the river pattern, the crack propagated circumferentially outward in both directions from the bracket leg. No crack arrest marks were noted on the fracture surface, and the edges of the wall did not contain any shear lips.[9] No corrosion damage or gouge marks were found on the inside surface in the general area of the fracture origin.

For a detailed examination, a portion of the fracture that included the fracture origin area, welded joint, void, and bracket leg was excised from the tank car. A dark region that was as wide as the hole in the eduction pipe guide (about 4 inches in diameter) was noted on the fracture surface between the legs of the bracket. The dark region was also excised. After the two excised fractures had been cleaned, they were examined with a scanning electron microscope, revealing cleavage features typical of a brittle overstress fracture. No preexisting or progressive fracture features were noted.

Another saw cut was made through the bracket leg-to-tank wall weld joint in an area 1 inch behind the fracture. A visual examination of the saw cut surface revealed that the root of the weld contained a void filled with weld slag, which was approximately the same size as the void at the fracture origin. The polished and etched wall of the tank car contained a microstructure of ferrite and pearlite, which is typical of a low carbon steel. Rockwell hardness testing of the etched surface in the area of the wall, weld, and bracket leg produced average hardness values, which converted to a tensile strength that was within the specified range for TC-128 steel. Additionally, a chemical analysis of the TC-128 steel from tank car TEAX 3417 indicated that the concentrations of carbon, manganese, sulfur, and silicon were all within their specified ranges.

To assess the ductile-to-brittle properties of the TC-128 steel from tank car TEAX 3417 and to determine the temperature at which the failure occurred, Safety Board investigators considered the standards for Charpy impact tests in the AAR specifications for TC-128 steel and then had an independent laboratory conduct Charpy impact tests on specimens from the tank car. (The appendix discusses the Charpy impact tests and how they can be used to evaluate ductile-to-brittle transition temperatures.) The January 1, 1970, version of the AAR specification[10] for TC-128 steel indicated that the Charpy impact properties were to be specified in the DOT tank car specification. (Neither the AAR nor the FRA had records to indicate what the impact properties were.) However, the 1994 version of the AAR specifications for TC-128 steel did specify minimum energy absorption standards for Charpy impact testing: a minimum average of 15 foot-pounds (ft-lbs) for three specimens and a minimum of 10 ft-lbs for each single specimen at a test

[9] A crack arrest mark is a "step" on the fracture surface and indicates an intermittent stopping point during fracture propagation. A shear lip, a narrow slanting ridge at the edge of a fracture surface, is present in materials that exhibit ductility but is absent in brittle materials.

[10] AAR Specification M128.00, *Specification for High Strength Carbon Manganese Steel Plates for Tank Cars - AAR TC-128.*

temperature of -50 °F and in the longitudinal direction of rolling.[11] A portion of the tank wall from the tank car was cut out and sent to an independent laboratory, where Charpy V-notch specimens were manufactured[12] from the longitudinal direction and tested at -50 °F. The average impact energy for the three specimens was calculated to be 8 ft-lbs. Only one of the three specimens attained the minimum absorbed impact energy of 10 ft-lbs for a single specimen.

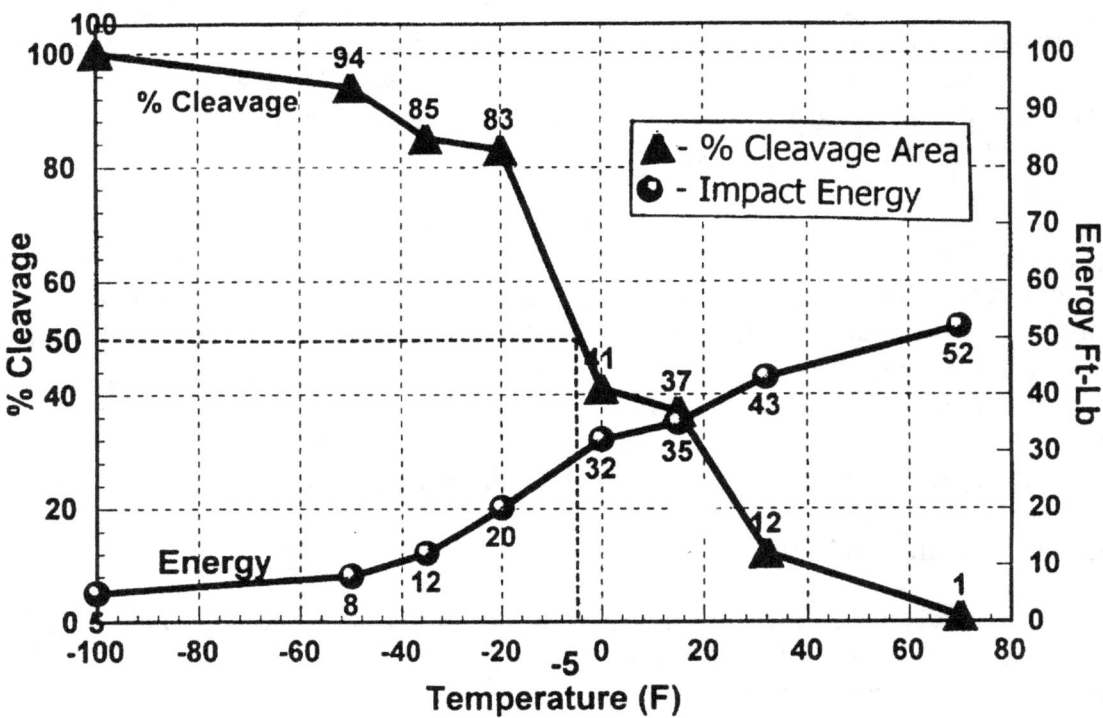

**Figure 4. Percentage of fracture surface exhibiting brittle features versus test temperature for Charpy test specimens**

---

[11]Under the AAR specification, the TC-128 steel had to satisfy these energy absorption standards only when the tank car was designated for cold-temperature service. Although the AAR specification did not define cold-temperature service, an FRA representative stated that the term was applied to products, such as carbon dioxide, vinyl fluoride, and hydrogen chloride, that were loaded at temperatures below -20 F. Safety Board investigators found no records to indicate that tank car TEAX 3417 was manufactured for cold-temperature service.

[12]Manufactured according to American Society for Testing and Materials (ASTM) E23, *Standard Methods for Notched Bar Impact Testing of Metallic Materials*, published April 1986. Specimens were standard size (10 mm x 10 mm), type A, notch facing forward or to the aft side of the tank car.

To determine the temperature at which the tank car wall fractured and the ductile-to-brittle transition temperature of the steel from the tank car, Safety Board investigators had the same independent laboratory manufacture additional Charpy specimens from the tank car's bottom wall and test each of three specimens at 70 °F, 32 °F, 15 °F, 0 °F, -20 °F, -35 °F, and -100 °F. A visual examination of the fractures from all the Charpy specimens disclosed progressively larger areas with features of cleavage (brittle) fracture as the testing temperature was lowered. The fracture surface of the specimen tested at -100 °F contained brittle features throughout the fracture area and no shear lips at the edges of the specimen, and was most similar to the fracture surface from tank car TEAX 3417. In contrast, the fractures tested at all the other intermediate temperatures contained a combination of brittle and ductile dimple features and shear lips at the edges that were not observed on the fracture surface from tank car TEAX 3417.[13]

The percentage of the fracture surface exhibiting brittle features from the Charpy test specimens was plotted versus the test temperature. (See figure 4.) From this plot and the industry-accepted definitions of transition temperature described in the appendix, the fracture appearance transition temperature (FATT) with 50-percent cleavage for the TC-128 steel from the tank car was about -5 °F, and the shear area transition temperature for the TC-128 steel was about 30 °F. The nil ductility temperature, the temperature at which the TC-128 steel exhibited no ductility (was completely brittle), was between -50 °F and -100 °F.

# Purging Procedures and Standards

## Repair Facility Procedures

During the meeting on January 7, 1998, Bayou representatives described the company's nitrogen purging procedures. The procedures used to purge tank car TEAX 3417 were not put in a written form until Safety Board investigators requested a copy of them. Under Bayou's purging procedures, an uninsulated copper tube connected the vapor outlet on a DOT specification 4L nitrogen cylinder to the liquid eduction line on the tank car. According to a plant manager with Taylor Wharton (the manufacturer of the nitrogen cylinder used at Bayou), the internal pressure of the nitrogen in the cylinder ranged from 123 psig to 237 psig. Based on the thermodynamic properties of nitrogen, the maximum temperature for the compressed liquefied nitrogen would have been -274 °F at 123 psig and -256 °F at 237 psig.[14] Because of the higher pressure in the cylinder, nitrogen flows from the cylinder through the hose and liquid eduction line to the

---

[13]The cleavage and ductile features were verified by scanning electron microscope examination.

[14]At atmospheric pressure, the boiling point of nitrogen is -320 °F. As pressure over the liquid nitrogen increases, the boiling point also increases.

bottom of the tank car. The flow of nitrogen is turned off at the cylinder after 20 minutes, and the vapor mixture in the tank vents through the tank car's safety relief valve. The oxygen content of the escaping gas mixture is tested, and the cycle is repeated until the oxygen concentration reaches a customer-requested level, usually around 2 percent. Bayou employees could not remember purging tank car TEAX 3417.

According to National Welding Supply, a vendor that fills and services the nitrogen cylinders for Bayou, each cylinder has an internal vaporizer that warms and converts the liquefied nitrogen to gas as the nitrogen flows from the cylinder. According to the same vendor, as the rate of flow from the cylinder increases, the capacity of the internal vaporizer to warm the nitrogen is exceeded, resulting in cryogenic gaseous or liquid nitrogen flowing from the cylinder into the tank car. Cylinder manufacturers therefore recommend the use of external heat exchangers (cryogenic vaporizers) between the cylinder and the tank car to ensure that the nitrogen is sufficiently warmed at increased flow rates. Bayou did not use an external heat exchanger in the purging of tank car TEAX 3417.

Shortly after January 1998, when Safety Board investigators examined tank car TEAX 3417, Bayou modified its nitrogen purging procedures. Under what Bayou designated as its Phase II system, nitrogen no longer flowed from the nitrogen cylinder directly to the tank car, but instead flowed from the cylinder into a "secondary vapor tank" and then to the tank car. Also, the nitrogen hose was connected to the tank car's vapor line, rather than to the liquid eduction line, so that nitrogen was injected through the vapor line, which extended only a short distance into the tank. Under Bayou's present system, its Phase III system, an external heat exchanger equipped with two high-pressure relief valves and a low-temperature sensor replaced the secondary vapor tank. (Bayou confirmed that the external heat exchanger was delivered on November 12, 1998, and was operational on November 13.) Consequently, nitrogen now flows from the nitrogen cylinder through the external heat exchanger, the pressure relief valves, and the low-temperature sensor before it is injected into the tank car.

## Compressed Gas Association Procedures

In 1992, the Compressed Gas Association, Inc., (CGA) recommended procedures for purging tank cars with nitrogen.[15] The CGA noted that introducing liquid nitrogen into tank cars not specifically designed for cryogenic service can result in tank failure caused by a phenomenon known as thermal shock (very rapid and severe cooling of the tank). To prevent thermal shock, the CGA recommended using cryogenic vaporizers that are external to the nitrogen cylinders to warm the liquid nitrogen to a gas with a temperature of not less than -20 °F before the nitrogen is injected into a tank car. The CGA also recommended using a low-temperature protection device that stops the flow of nitrogen to the tank car if the temperature of the nitrogen falls below -20 °F. Under the CGA

---

[15]*Recommended Procedures for Nitrogen Purging of Tank Cars*, CGA Pamphlet P-16-1992.

procedures, the nitrogen can be injected through either the liquid eduction line or the vapor line of a tank car.

The CGA was unable to locate any technical information that might have been used in developing its nitrogen purging procedures. Safety Board investigators contacted some of the CGA committee members who had developed the procedures, including those members who represented tank car manufacturers and owners, in an effort to determine how the -20 °F temperature threshold had been derived. The members contacted could not recall whether the threshold was based on any specific testing or engineering analysis. They indicated that the threshold had most probably been recommended by tank car experts who were drawing on their collective experience and that the committee had, therefore, adopted the recommendation. All the committee members contacted stated they were no longer able to locate any meeting notes that would verify their accounts of how the threshold had been determined.

To determine whether the CGA procedures had been adopted and implemented at other tank car repair shops, Safety Board investigators contacted four other companies that own and operate tank car repair shops. A comparison of the purging procedures employed by the CGA, Bayou, and the four companies are provided in table 1.

| Table 1: Comparison of nitrogen purging procedures | |
|---|---|
| CGA | Recommends using external cryogenic vaporizers able to convert liquid nitrogen into a -20 °F vapor. Recommends using a low-temperature shut off device set at -20 °F. |
| Bayou | Lets nitrogen flow directly from a nitrogen source vapor outlet to a tank car liquid line. Does not use external vaporizers or a low-temperature protection device. |
| American Railcar Industries | Uses external cryogenic vaporizers but sets the low-temperature protection device at -40 °F. |
| General American Transportation | Bases its system on the CGA purging procedures. |
| Trinity | Uses external cryogenic vaporizers but does not monitor vapor temperatures. |
| Union Tank Car | Bases its system on the CGA purging procedures. |

## Tank Car Failures Resulting from Nitrogen Purging

Both the AAR and the FRA have documented tank car failures that resulted from thermal shock caused by nitrogen purging. In 1982, following the failure of a tank car in Ontario, Canada, the AAR issued an industry circular letter warning of the dangers associated with the use of nitrogen during purging and padding.[16] In 1994, due to a failure in Mississippi attributed to the injection of nitrogen, the FRA urged the AAR, the Railway Progress Institute, and the Chemical Manufacturers Association to remind their members that nitrogen purging can cause thermal shock. The FRA also forwarded copies of the CGA procedures to these organizations for distribution to their members.

Additionally, the FRA investigated the 1997 failure of tank car MCTX 9009, a DOT specification 112J340W tank car constructed in 1969 with TC-128 steel. The tank car sustained a circumferential fracture under the eduction pipe guide bracket. The FRA determined, through its investigation and a metallurgical analysis, that the crack occurred after the tank car was purged with nitrogen by a shop that failed to use an external vaporizer and a low-temperature sensor.

## AAR/FRA Actions

As a result of the failure of tank car TEAX 3417, the AAR Tank Car Committee established a working group in March 1998 to consider amending the AAR tank car manual[17] regarding the purging of tank cars with nitrogen. At the July 1998 meeting of the Tank Car Committee, the working group approved a revision. Under the revision, each facility that purges a tank car with nitrogen or other cryogenic liquid "shall ensure against the introduction of liquid into the tank through a written procedure, training, supervision, and appropriate equipment." The revision also recommends the nitrogen purging procedures in the latest edition of the CGA purging procedures. The revision, however, does not address the adequacy of the -20 °F temperature threshold.

The FRA representative who was a member of the working group and assisted in developing the proposed revision told Safety Board investigators that the revision had been approved and that the use of written procedures for purging will become an industry standard upon publication of the 1999 edition of the tank car manual. He said that as a result, purging procedures will be subject to quality assurance audits of tank car facilities by both the AAR and the FRA.

---

[16] A procedure of injecting nitrogen or another inert gas into the vapor space of a loaded tank car with a product that for reasons of product purity or reactivity should not be exposed to air and moisture in the vapor space.

[17] *Manual of Standards and Recommended Practices, Section C-Part III - Specifications for Tank Cars, Specification M-1002*, AAR, Washington, DC.

# Loading of Liquefied Gases

## Clark's Procedures

Tank car TEAX 3417 was loaded from tank car PLMX 36600, which Clark was using for storage. Clark's practice is to reduce the internal pressure as much as possible in the receiving tank car when loading it from either a fixed storage tank or a tank car used for storage (storage tank car). When a liquefied compressed gas is transferred from a fixed storage tank, Clark uses a pump. When the liquefied compressed gas is transferred from a storage tank car, Clark injects the storage tank car with nitrogen so that the pressure differential between the storage tank car and the receiving tank car is enough to complete the transfer. Because of the low pressure in the receiving tank car, when it is injected with liquefied compressed gas, the gas "flashes," or vaporizes, almost instantaneously.

## Industry Evaluations of Flash Loading

In March 1982, following the failure of two DOT 105A400W specification tank cars (LEYX 636 and LEYX 639), the AAR, under the auspices of its Tank Car Committee, charged a task force to "investigate the compatibility of spray filling of liquefied compressed gases or anhydrous ammonia with tank car tank design." Although each of the two failures was attributed to the presence of existing cracks that had been caused during nitrogen purging, the task force also acknowledged the potential for "undesirably low shell temperatures" if the tank has been evacuated to a low internal pressure before compressed liquefied gases are loaded, a practice known as "flash loading." To avoid such a low shell temperature, the task force recommended that the compressed liquefied gas be loaded by spraying it at a high flow rate through the vapor valve on the tank car during the "initial fill period." According to AAR records, Amoco was to conduct tests to document how the temperature of a tank shell is affected by spray loading. In response to inquiries from Safety Board investigators about the tests, an Amoco representative stated that they had been done, but that the company had not kept records of the results. The representative was unable to recall the results.

According to the AAR, in 1996 approximately 168,100 tank cars were loaded with LPG and 65,000 tank cars with anhydrous ammonia. In 1997, approximately 185,500 and 66,200 tank cars, respectively, were loaded with LPG and anhydrous ammonia. The figures for both years include the tank cars that were loaded in both the United States and Canada. Neither the FRA nor the AAR could identify any known instances of thermal shock failure attributed solely to flash loading of LPG or anhydrous ammonia.

# Analysis

## Failure of Tank Car

On the basis of a metallurgical examination, the crack and the resulting release of the propane/propylene were caused by a brittle overstress fracture that originated from a void (creating a stress riser) at the root of a weld joint between the A-end eduction pipe bracket and the inside surface of the tank wall. The fact that no crack arrest marks or fatigue cracks were found in the wall of the tank car tank suggests that the fracture was caused by a single event. The wall of the tank car was manufactured to the correct chemical composition, and the tensile strength (converted from hardness testing) was within the specified range.

The features of the fracture surface from the tank car were most similar to those of the fracture surface from the Charpy V-notch specimen tested at -100 °F. Both surfaces contained brittle features, and the edges of the fractures did not have any shear lips. Further, the results of the Charpy V-notch testing indicate that brittle (cleavage) features on 100 percent of the fracture surface were attained only for the specimen tested at -100 °F. In contrast, the fracture surfaces of the Charpy V-notch specimens tested at -50 °F and above contained ductile dimple features that were not observed on the crack surfaces from the tank car. These results demonstrate that at some temperature between -50 °F and -100 °F the fracture surfaces of the steel from tank car TEAX 3417 would change from exhibiting a combination of ductile and brittle features to exhibiting 100 percent brittle features. Because the fracture surface from the tank car exhibited only brittle features and contained no crack arrest marks, the Safety Board concludes that the tank car cracked as a single event as a result of thermal shock that had locally cooled the tank car to less than -50 °F.

## Event Leading to Thermal Shock Failure

The event that caused the thermal shock, according to the metallurgical analysis, would have cooled the tank wall beneath the eduction line to less than -50 °F. Between October 11, when the tank car was hydrostatically tested at Bayou, and November 22, when the leak was discovered, the tank car was not exposed to ambient air temperatures below 20 °F. Therefore, ambient air temperatures were not a factor in the failure of the tank car.

Rather, the fact that the crack was immediately below the eduction line, which is used for loading and offloading, suggests that the event causing the thermal shock probably occurred during the loading or injecting of very cold material into the tank. The material touched and cooled the tank below the eduction pipe, resulting in thermal shock. Safety

Board investigators identified two events that could have caused the thermal shock: the purging of the tank car at Bayou and the loading of the product at Clark. Both events involved injecting a compressed liquefied gas into the tank car with little or no internal pressure. Under such conditions, any liquefied nitrogen and liquefied propane/propylene "flashes," or vaporizes, almost instantaneously when it is discharged from the eduction line. As a result, any structure in contact with the vaporizing liquefied gas cools rapidly. The extent of cooling depends upon the discharge temperatures of the nitrogen or the propane/propylene mixture from the eduction line and upon the boiling points of the materials.

Based upon an analysis of these two events, the Safety Board concludes that the crack is more likely to have occurred during the purging of the tank car. Because Bayou did not use external cryogenic vaporizers and a low-temperature protection device, the tank car was probably injected with cryogenic nitrogen, including entrained liquid nitrogen. The liquefied nitrogen in the cylinder was under an internal pressure between 123 psig and 237 psig and would, therefore, have had a maximum temperature between -256 °F and -274 °F. Consequently, if the liquefied nitrogen had any contact with the tank shell directly below the eduction line, the result would have been the very rapid super cooling of the tank shell in the area of contact to less than the -50 °F threshold. In comparison, the loading temperature of the liquefied compressed propane/propylene mixture was 41 °F, well above the maximum temperature of -50 °F at which the thermal shock failure occurred. Although flash loading a liquefied compressed propane/propylene mixture also cools the tank shell in the area below the eduction pipe, the loading temperature of the mixture and the boiling points of the components (-44 °F for propane and -54 °F for propylene) would have resulted in exposing the tank to vaporizing liquid at or above -54 °F and would not have caused as extreme cooling of the tank as would occur with cryogenic materials, such as liquefied nitrogen.

Further, the AAR and the FRA have documented other instances of thermal shock failures following nitrogen purging of tank cars. The common factor in the incidents is the failure to employ external cryogenic vaporizers and a low-temperature protection device. The 1997 failure of tank car MTCX 9009 bears a striking resemblance to that of tank car TEAX 3417. Both tank cars had circumferential cracks that originated in pipe guide bracket welds. The FRA determined that tank car MTCX 9009 failed because safeguards were not used in the nitrogen purging system, as, apparently, they were not used with tank car TEAX 3417. Based on the similarities between the failure of tank car TEAX 3417 and the documented cases of other tank cars failing during nitrogen purging, particularly the case of tank car MTCX 9009, it is probable that tank car TEAX 3417 failed while it was being purged at Bayou.

In contrast to the examples of tank car failures following nitrogen purging, the FRA has not recorded any instances of tank car failures due solely to low temperatures created by flash loading of liquefied compressed gas. Although the AAR has documented two instances of tank cars failing while they were being loaded, both failures were attributed to preexisting cracks caused by nitrogen purging. The AAR has not documented a tank car failure attributed solely to flash loading. Additionally, of the approximately 354,000 tank cars loaded with LPG and 131,000 tank cars loaded with anhydrous ammonia during 1996 and 1997, none were reported to the AAR as failing because their tank shells had been exposed to low temperatures during flash loading. Consequently, there is no indication of a pattern of tank car failures due to flash loading of liquefied compressed gases. Consequently, the Safety Board believes that the cause of the thermal shock failure of tank car TEAX 3417 was the failure of Bayou to utilize sufficient safeguards to ensure that the nitrogen was properly warmed before it was injected into the tank car during nitrogen purging.

Since the accident, Bayou has taken positive and appropriate actions to improve its nitrogen purging procedures. The Phase III system that Bayou recently installed incorporates the use of an external heat exchanger and low-temperature sensors, two essential safeguards that have been recommended by the CGA. The Safety Board believes that with these improvements, Bayou has minimized the chance of having another tank car fail because of thermal shock caused by nitrogen purging.

## Nitrogen Purging Procedures

External cryogenic vaporizers and low-temperature protection devices, as recommended in the CGA nitrogen purging procedures, are safeguards that tank car facilities must employ during nitrogen purging to ensure that the extremely cold nitrogen gas or liquid does not come into contact with the tank car shell. Nevertheless, the minimum injection temperature of -20 °F specified in the CGA standard has not been verified through scientific tests or engineering analysis.

Based on the definition of the ductile-to-brittle transition temperature as the temperature at which 50 percent of a fracture exhibits shear features and 50 percent exhibits cleavage features, the transition temperature for the TC-128 steel in tank car TEAX 3417 would be about -5 °F. Because TC-128 steel has been the most commonly used grade of steel in the construction of tank cars over the past 20 or more years, the likelihood is strong that the steel found throughout the tank car fleet will have comparable ductile-to-brittle transition temperatures. Further, the Safety Board is concerned that other grades of tank car steels may also have high transition temperatures. The Safety Board determined in its investigation of the 1996 failure of tank car GATX 92414 in Sweetwater, Tennessee,[18] that the transition temperature of the ASTM A-515-70 grade steel[19] from GATX 92414 was 30 °F. Consequently, because the CGA's minimum injection temperature is not supported by an engineering evaluation and because grades of steel used in tank cars have high ductile-to-brittle transition temperatures, the Safety

---

[18]Hazardous Materials Accident Brief--*Tank Car Failure and Release of Flammable and Toxic Liquid, Sweetwater, Tennessee, February 7, 1996*. National Transportation Safety Board. April 20, 1998.

[19]The AAR prohibited the use of this grade of steel in 1987 and replaced it with ASTM A-516-70 steel.

Board concludes that the current CGA procedures for nitrogen purging of railroad tank cars do not adequately protect the tank cars from brittle failure.

Therefore, the Safety Board believes that the CGA, with the assistance of the FRA and the AAR, should revise the CGA's *Recommended Procedures for Nitrogen Purging of Tank Cars* to specify a minimum threshold temperature for nitrogen that is based on an engineering analysis of ductile-to-brittle transition temperatures of tank car steels.

# Findings

## Conclusions

1. Tank car TEAX 3417 cracked as a single event as a result of thermal shock that had locally cooled the tank car to less than -50 °F.

2. Tank car TEAX 3417 probably cracked during the nitrogen purging conducted at Bayou Railcar Services, Inc., on October 17, 1997.

3. Because the minimum injection temperature recommended by the Compressed Gas Association, Inc., is not supported by an engineering evaluation and because grades of steel in the tank car fleet have been found to have high ductile-to-brittle transition temperatures, the procedures of the Compressed Gas Association, Inc., for nitrogen purging of railroad tank cars do not adequately protect the tank cars from brittle failure.

## Probable Cause

The National Transportation Safety Board determines that the probable cause of the thermal shock failure and the subsequent leakage of product from tank car TEAX 3417 was the failure of Bayou Railcar Services, Inc., to utilize sufficient safeguards to ensure that the nitrogen was properly warmed before it was injected into the tank car during nitrogen purging.

# Recommendations

As a result of this accident, the National Transportation Safety Board makes the following safety recommendations:

**--to the Compressed Gas Association, Inc.:**

In cooperation with the Federal Railroad Administration and the Association of American Railroads, revise its *Recommended Procedures for Nitrogen Purging of Tank Cars* to specify a minimum threshold temperature for nitrogen that is based on an engineering analysis of ductile-to-brittle transition temperatures of tank car steels. (R-98-70)

**--to the Federal Railroad Administration:**

Cooperate with the Compressed Gas Association, Inc., in its efforts to revise its *Recommended Procedures for Nitrogen Purging of Tank Cars* to specify a minimum threshold temperature for nitrogen that is based on an engineering analysis of ductile-to-brittle transition temperatures of tank car steels. (R-98-71)

**--to the Association of American Railroads:**

Cooperate with the Compressed Gas Association, Inc., in its efforts to revise its *Recommended Procedures for Nitrogen Purging of Tank Cars* to specify a minimum threshold temperature for nitrogen that is based on an engineering analysis of ductile-to-brittle transition temperatures of tank car steels. (R-98-72)

**BY THE NATIONAL TRANSPORTATION SAFETY BOARD**

**JAMES E. HALL**
Chairman

**JOHN A. HAMMERSCHMIDT**
Member

**ROBERT T. FRANCIS II**
Vice Chairman

**JOHN J. GOGLIA**
Member

**GEORGE W. BLACK, JR.**
Member

December 1, 1998

# Appendix: Transition from Ductile-to-Brittle Behavior in Steels

## Charpy Testing

Charpy testing assists designers in selecting materials that resist brittle fractures and normally involves a single impact test in which a notched specimen is supported at both ends as a single beam and is then broken by a falling pendulum. The amount of brittle (cleavage) fracture, expressed in percent, for one Charpy specimen at each testing temperature, is determined by measuring, from a photograph of a particular specimen, the fracture area with cleavage features and then dividing this area by the total fracture area of the specimen.

Charpy testing results generally do not vary among steels manufactured from the same lot. However, the results can vary significantly for the same grades of steel manufactured from different lots. The transition temperatures for different lots may vary by several degrees. Charpy testing is very sensitive to the variables in the processing of the steel, such as heat treatment, grain size and orientation, and alloying elements.

## Transition Temperatures

Structural steels tend to go through a transition from ductile behavior at high temperatures to brittle (cleavage) fracture at lower temperatures. Transition temperature is often defined as the temperature needed to attain a given fracture appearance (that is, 50 percent brittle fracture), an arbitrary level of fracture energy (that is, 15 ft-lb), or some specified level of ductility (that is, 15 millimeters of lateral expansion). With many steel alloys, the transition does not occur at a specific temperature, but over a temperature range. Consequently, it is often difficult to define ductile-to-brittle transition temperature accurately.

There are several industry-accepted definitions for transition temperature. The FATT is based on the physical appearance and features of the fracture. A common definition for the FATT is the temperature at which the fracture surface from a Charpy test specimen has 50 percent cleavage (brittle) and 50 percent shear (ductile) features. According to industry-accepted practice, materials that are exposed to service conditions below this transition temperature will exhibit brittle fracture. Separations of materials below this transition temperature are defined as brittle fractures. A more conservative approach, based upon "shear area" defines the transition temperature as the temperature at which a fracture has shear features over 85 percent of the fracture surface. Stress analysts in the pipeline industry consider embrittlement in steel to occur below the shear area transition temperature. A third definition of the transition temperature is the nil ductility temperature, or the temperature at which the fracture surface exhibits 100-percent cleavage.

A transition temperature curve for a specific material can also be obtained by preparing a graph plotting the test temperature at which the Charpy specimen is fractured versus the impact energy that is required to fracture the specimen. The curve does not show the transition temperature directly, but must be derived from the constructed curve. The curve will typically have an "upper" and "lower" shelf where the fracture energy remains nearly constant relative to temperature. The average of the temperatures at which the upper and lower shelves occur is often defined as the transition temperature.

# Acronyms and Abbreviations

AAR: Association of American Railroads

Bayou: Bayou Railcar Services, Inc.

CFR: *Code of Federal Regulations*

CGA: Compressed Gas Association, Inc.

Clark: Clark Refining and Marketing

DOT: U.S. Department of Transportation

FATT: fracture appearance transition temperature

FRA: Federal Railroad Administration

ft-lb: foot-pound

LPG: liquefied petroleum gas

psig: pounds per square inch, gauge

TEI: Transportation Equipment, Inc.

Trinity: Trinity Industries, Inc.

www.ingramcontent.com/pod-product-compliance
Lightning Source LLC
Chambersburg PA
CBHW081819170526
45167CB00008B/3470